DYLIFE

DAVID E. BICK
B.Sc., M.I.Mech.E.

Published by THE POUND HOUSE, Newent, Glos.
Printed by Albert E. Smith (Printers) Ltd., Gloucester

Richard Cobden, M.P. 1804-1865   (British Museum)

CONTENTS

Introduction 5
General History.. 6
Machinery and Industrial Archaeology 19
Output Statistics 30
Sources and Acknowledgements 31
Glossary 32

MAPS AND DIAGRAMS

Dylife 4
Esgairgaled and Llechwedd Ddu Section, 1849 8
The Esgairgaled Area in the 1880's 10
Lode Outcrops and Watercourses 11
A Section of the Dylife Lode 18
An Account of the Lead Mine in Delivie Mountain 19
A Survey of Five Adits 29

PLATES

Richard Cobden, M.P... .. .. Frontispiece
Llechwedd Ddu Engine Shaft 9
Dylife about a hundred years ago 13
Dylife Post Office 21
Balance bob gudgeon at Boundary Shaft 22
Wheelpit of the Red Wheel 23
The Star Inn in 1950 24
Captain Edward Williams 25
Footway Shaft 27
Revealing the Llechwedd Ddu cage 27
John Bright, M.P. 28

Dylife in the 1880's.

Note the remarkable mile-long cable running diagonally across the map.

Ordnance Survey.

4

INTRODUCTION

Until the close of the last century, one of the most important industries in Wales was metal mining and mellowed remains are to be found in almost every corner of the Principality. Its history has been told in a general way but I believe the need is growing for a more specific coverage devoted to individual mines, with guidance for those wishing to visit the sites and to explore the physical remnants for themselves. Indeed, only in this way can we really appreciate the difficulties facing the 'old men' and gain an insight into a forgotten world.

High in the mountains between Llanidloes and Machynlleth lies Dylife, and in choosing it as my first subject I am conscious that here was neither the best known nor the most productive of lead mines. Nevertheless, it achieved greatness in spite of a geographical isolation which in those days was almost complete.

By the 1860's some of the most modern mining machinery in Britain found application at Dylife and its owners, who included Cobden and Bright the social reformers, were rewarded with huge profits. There are few sites more worthy of study, and none where I have spent more pleasurable times, both above ground and below.

Finally a word about exploration. Footpaths and other rights of way are marked on the latest 1:50,000 Ordnance map, sheet 135 ; it is hardly necessary to add that great care should be taken to close gates and to avoid damage to fences. In addition, permission should always be obtained prior to entering private property. Underground adventures are best avoided unless accompanied by experienced guides, whilst open shafts and workings are especially treacherous. Like magnets, their attraction grows as the gap diminishes !

GENERAL HISTORY

Mining at Dylife, as in many other regions of Mid-Wales, dates from Roman or even earlier times. Antiquarians were studying the area as long ago as 1857 when a level 150 fathoms long had been revealed 'only ten inches wide at the bottom, then bulging out to make room for the shoulders in the shape of a coffin set on its smaller end'. Remains of a smelting furnace have also been found on the north side of Dylife hill and attributed to the Romans ; unfortunately details of the site are lacking.

The lead, zinc and copper ores were deposited in mineral lodes or veins which occur abundantly in the Silurian rocks surrounding Plynlimmon, and represent the gradual filling of almost vertical fissures by the action of mineralized solutions. The workings at Dylife centred on the Esgairgaled, Llechwedd Ddu and Dylife lodes, and these veins constitute the fan-like eastern extremity of a very powerful fracture that ranges through the mountains into Cardiganshire. The Llechwedd Ddu lode was a relatively late discovery, though one of prime importance.

In the 1720's a Mr. Edward Harley was working a mine at 'Eskergallid', and 50 years later the Flintshire firm of Bowdler Vickers were active in the same area, but the mines did not reveal their true potential until the 19th century, following the general upsurge of metal mining in Britain. This era may conveniently be divided into several main headings as below.

Williams & Pughe	c. 1818—1858
Cobden, Bright & Co.	1858—1873
Dylife Lead Mining Co.	1873—1876
Great Dylife Companies	1876—1884
Blaen Twymyn	1886—1891

By 1818 Hugh Williams of Machynlleth had taken over the lease and about this time a new lode (no doubt the Llechwedd Ddu) quickly proved extremely rich. Williams also worked the Dylife lode high on the hill south of the Llanidloes—Machynlleth coach road, which had already received much attention from the 'old men'. John Pughe of Aberdovey joined forces with Williams and by 1845, 70 tons of lead ore and 10-12 tons of copper ore were being returned monthly.

Perhaps the earliest surviving details of underground workings date from 1849, when a report on the several separate mines constituting Dylife was prepared by Matthew Francis, a Cornish mining engineer resident in Cardiganshire.* From this we learn that the Dylife lode had not

*Francis came to Wales in 1834 as agent for the prominent firm of mining engineers, John Taylor & Sons. After his dismissal in 1842 he turned to consulting and promotional work, during the course of which he found it prudent to err on the side of optimism. However, in the case of Dylife, Francis's forecasts were borne out almost to the letter.

been worked for some years ; the main activity centred on the Llechwedd Ddu lode, where the Engine (pumping) shaft had reached the 45 fm level. It had been sunk near a barren zone, but to the eastward and westward the lode was found productive 'for a length of 80 fathoms, in many places yielding 3 or 4 tons of lead with a few cwts of copper to the cubic fathom. . . . I would advise,' wrote Francis, 'the 35 or 45 fm level should be continued Eastwards for the purpose of discovery, it being very improbable that the Lode should carry such fine bodies of ore as those already laid open without there being others of similar quality in the neighbouring ground'. Francis went on to observe that in going eastward the lode split into three branches.

The western end of the Esgairgaled lode was known as Gwaith Gwyn (White Works) and here a crosscut level had been driven north from Llechwedd Ddu engine shaft at the 25 fm level under old and exhausted surface workings ; Francis recommended that it should be completed, but on carrying this out, the lode unhappily proved so impoverished that the project became terminated until about 1861, when a short drivage east ran into 18 ins width of solid galena. This soon expanded to fully a yard from which several thousand tons were raised. Subsequently by stripping down the walls of the lode west of the crosscut an equally rich body of ore was revealed, the two deposits in fact very much mirroring those of the parallel Llechwedd Ddu lode.

However, to return to Francis's report. At this time (1849) old men's workings extended 200 fathoms through the Esgairgaled and Pencerrig properties, pointing to a large potential in depth. He advised further sinking the Esgairgaled shaft which was then only 20 fathoms deep, and enlarging and extending the old Level Goch (Red Level) on the supposed united Llechwedd Ddu and Esgairgaled lodes. Due to the rapid growth of the mines, available waterpower was becoming quite inadequate and Francis also suggested construction of several large reservoirs and other surface developments, as described later.

So far, nothing has been said about the great adit crosscut which intersected the Dylife lode 60 fathoms west of Old Engine shaft after a drivage of 1/3rd mile. It must have represented the labour of many years and was large enough to admit horses for pulling mine trams. Though the portal is close to Llechwedd Ddu engine shaft, this was not the original entrance, for the crosscut began from the Llechwedd Ddu adit which itself had been driven on the lode, commencing some distance to the east. Prior to 1849 in order to make a more convenient access, a short inclined plane was constructed from the crosscut close to its intersection with the Llechwedd Ddu adit, to emerge near the engine shaft. The entrance to this incline has an impressive cavernous appearance.

The significance of so interesting a feature (which with caution can still be explored) has long been overlooked since almost a century ago the Llechwedd Ddu lode was said to have been discovered in driving the Dylife crosscut adit.

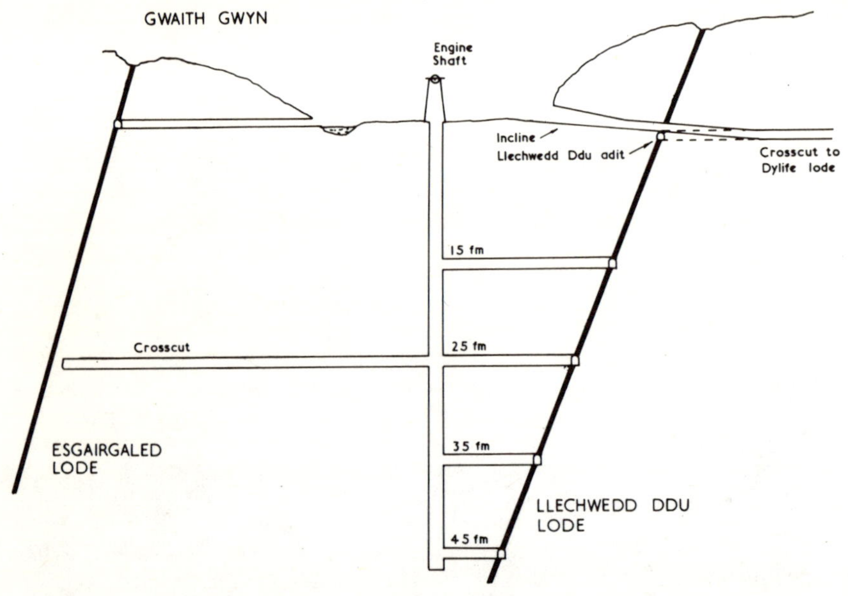

Esgairgaled and Llechwedd Ddu section, 1849.
Note the inclined plane and uncompleted crosscut to the Esgairgaled lode.
A plan of the inclined plane area is given on page 29.

By 1851 output was expanding rapidly, exceeding 1000 tons of lead ore, and in April 1852 Williams & Pughe went to the expense of entertaining their 300 men, women and children employees with tea, coffee and cold meat on the occasion of the lessor's (Sir Watkyn Williams Wynne) marriage. Intoxicating drinks were excluded but nevertheless this rural festival 'resounded with hilarity and song.'

In common with most metal mines, Dylife went through several bad patches where less perseverance would have led to premature closure. One of these was in 1855 when production fell to only 192 tons, by far the lowest for many years. Fortunately in June 1856 Jane Williams (widow of Hugh Williams who had died three years earlier) was able to record in a letter to Francis 'I went to the works on Friday last and was glad to find the prospects brightening. The lost sheep at Llechwedd Ddu has been found, a very good bunch of ore having been cut in the 75 fathom level . . .'.

Llechwedd Ddu Engine Shaft. *The uppermost portion of 600 ft. of pump rods makes a striking silhouette. Behind is the entrance to the great Dylife crosscut adit, and above can be seen outcrop workings on the Llechwedd Ddu lode.*

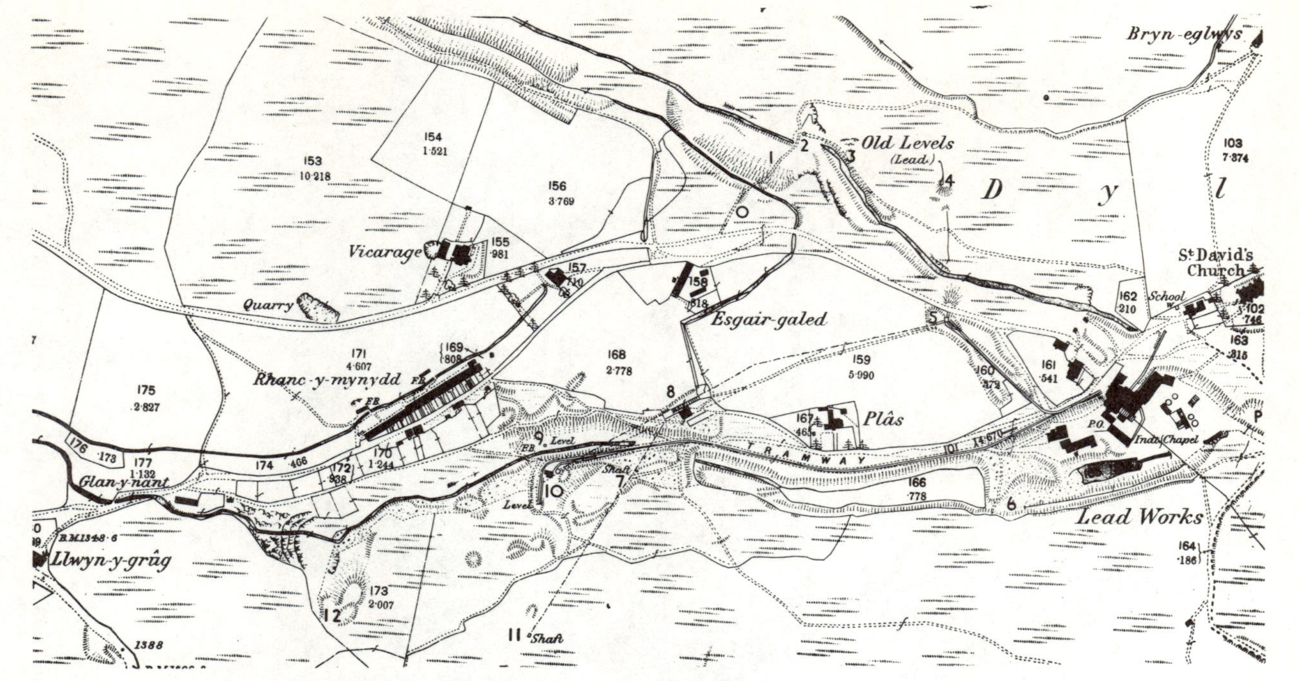

The Esgairgaled area in the 1880's. Each feature can still be traced.

25 in. Ordnance Survey.

0 Site of Black Wheel.
1 Footway shaft
2 Esgairgaled shaft & tunnel for stream
3 Pencerig deep adit
4 Pencerig shallow adit

5 Bradford's Shaft
6 Shaft on south lode
7 Footway shaft
8 63ft Red Wheel
9 Gwaith Gwyn adit

10 Llechwedd Ddu engine shaft and Dylife adit
11 Alfred's shaft
12 Junction of lodes
13 Level Goch

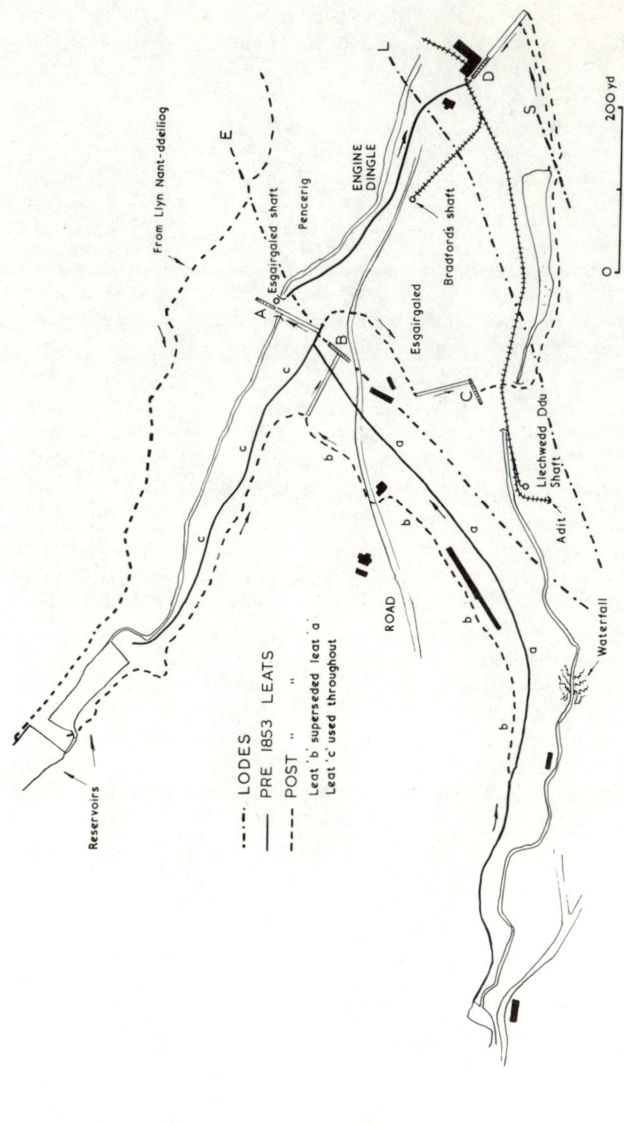

Lode outcrops and probable development of the watercourses.

A *First site of Black Wheel (the pit partly survives)*
B *Second site of Black Wheel*
C *63ft or Red Wheel (the pit survives)*
D *40ft crushing wheel*

E *Esgairgaled lode*
L *Llechwedd Ddu lode*
S *South lode, visible in stream*

11

The following year the mines were returning large profits, and far reaching changes were imminent. In 1840 Hugh Williams' daughter Catherine Anne had married Richard Cobden who was destined to become the great exponent of Free Trade. With John Bright, he worked tirelessly for the abolition of the corn laws that were literally starving the poorer classes to death. By 1858, control of the mine had passed to Mrs. Pughe and trustees of the late Hugh Williams including Richard Cobden himself.* Cobden was the son of a Sussex farmer and born in 1804.

These representatives however, disagreed amongst themselves, the dispute being settled by deciding to dispose of the mine in its entirety. In the interim period Dylife was managed by John Taylor & Sons who were Britain's foremost metal miners, and agents for the Wynnstay Estate.

Cobden directed the attention of his political friends including Bright and Milner Gibson to the mine, which was sold to a Manchester company formed by these men for the sum of £24,000. John Bright became chairman, and an agreement was reached with John Taylor & Sons to continue the management on their behalf.

At this time the practice of working non-ferrous metal mines lagged behind collieries, where from all large pits the coal and men were raised in cages. Metal mines perservered with the age-old kibble or at best, skips for winding ore, whilst an exhausting series of ladders perhaps a hundred or more fathoms in height faced the miners at the end of each shift. Such primitive, almost barbaric methods endured at nearly all Welsh mines till the end of the industry.

The Manchester company introduced prophetic changes by equipping Dylife in full colliery style, enabling ore to be trammed via cages directly to surface and hence to the dressing floors.

Within a few years an experienced inspector engaged upon an extensive survey of British mines was able to report 'I consider the whole of the winding machinery to be the most perfect to be found in the metallic mines visited.' Furthermore it was considered worthy of remark that Dylife was the sole mine in Wales and the North of England that provided changing rooms. In all other instances the men walked to and from their work—often wet through—in the clothes worn underground.

Thousands of pounds had been spent in these improvements which undoubtedly went far to reduce Dylife's working costs and to increase output. Whether John Taylor & Sons or the Manchester company (which may have included local colliery owners) initiated these revolutionary advances is an interesting question. But the new system was so out of character with Taylors' usual methods that we are pretty safe in assuming the famous firm of mining engineers were for once over-ruled. This is a part of Dylife's history we would like to know much more about.

*Just before his death, Williams reopened the Ceulan lead mine north of Dylife ; this also passed into the hands of Cobden & Bright.

Dylife about a hundred years ago, with dressing floors in the foreground.
(The photograph was taken from point P on the map, page 10, extreme right).

1. *Launders*   2. *The white building probably corresponds to that on page 21*.   3. *Chimney of horizontal 'puffer'*
4 *Headframe of Eastern Shaft, sunk on outcrop of Llechwedd Ddu lode*   5. *Headframe of Bradford's Shaft (130 fathoms deep)*   6. *Black Wheel, pumping and drawing from Esgairgaled Shaft.*

In the mine itself, one of the first aims of the new company was to make a determined attack on the Dylife lode, and this involved sinking a shaft (Boundary shaft) at the west end of the property adjacent to the Dyfngwm mine where good ore was being worked on the same lode.* Another important development was Bradford's shaft, conveniently close to the dressing floors and sunk in order to exploit the eastern workings on the Llechwedd Ddu lode, which in 1858 were down to the 85 fathom level at the engine shaft.

In July 1860 Bright wrote to Cobden 'Dylife is promising well and Bradford I think will allow us to have a dividend by September . . . I should like to spend a day with you on these glorious hills away from your diplomatic and my party tortures'. Later he added 'Bradford is our general managing director and we owe whatever success we meet to him'. At this period (January 1862) Bright stated that the Llechwedd Ddu lode was proving the great source of gain, though 'Esgairgaled has done pretty well'. 1862 was to prove Dylife's best year, with 2571 tons of lead ore sold, realizing profits of £1000 per month ; with the later exception of the Van lead mine near Llanidloes (for a while the biggest in Europe) this was to prove the greatest annual output of any mine in Mid-Wales.

When in 1863 evidence was being collected for a Government report into the working conditions of miners, there were 250 men underground of which the Dylife lode absorbed 38. 14 men were working the Esgairgaled lode at its eastern and western ends down to the 40 fathom level. In common with most other metal mines the ventilation was excessively bad in places though somewhat alleviated by blowing machines operated by boys turning a handle. The report also stated, in a matter of fact style, that 'in consequence of the scarcity of cottages, those in existence are very much overcrowded. In many cases lodgers are taken where there is not really room for the members of the family, giving rise to much overcrowding . . . As might be expected fever was raging to a great extent in the district'.

At this time Dylife's population amounted to about a thousand souls with three or four inns and a church and several chapels, together with a school. After 1864, opening of the Newtown-Machynlleth railway somewhat reduced transport costs since ore, coal, timber and general traffic could be carried via Llanbrynmair station, 6 miles to the north. Very probably the present road east from Dylife with its spectacular views of the Twymyn valley, was constructed for this purpose. Previously the port of Derwenlas on the Dovey 13 miles away was the chief route. All who have travelled the mountain road to Machynlleth will realise the daunting problems of transport before the days of motor vehicles, on roads little better than cart tracks.

---

*Dyfngwm was working hard at this time, and eventually reached 100 fathoms below adit. In 1867 Taylors' offered to buy a half share for £5,000, whilst Bright wanted to purchase the whole property.

When Richard Cobden died in April 1865, workings at Llechwedd Ddu were down to the 115 fathom level and the best ore had gone. After 1867 production went into a slow decline so that by 1872 it amounted to only a third of peak figures. Realising that Dylife was dying, Bright & Co. proceeded to dispose of it to the Dylife Lead Mining Company for no less than £73,000 at a time when according to a correspondent in the *Mining Journal* 'not an end in the mine would produce one cwt per fathom'.

Under the chairmanship of Offley Bohun Shore and management of Ralph Dean* the new company took over early in 1873 and as if in hopes that imitation would lead to equally successful results it laid out much money in improvements and underground development. Work centred on the Dylife lode ; Boundary shaft was sunk to 132 fathoms under adit, and a skip road installed. At Pencerig, the deep adit was abortively pushed eastward a great distance and a crosscut driven from the Llechwedd Ddu lode near the west end of the 105 fathom level to try the Esgairgaled lode in depth. None of these developments however, met with much success, in spite of glowing reports from such well known mining personalities as Walter Eddy (representing Sir W. W. Wynn), Captain Arthur Waters & Mr. Peter Watson.

Some of the heavy items of expenditure incurred during the company's first year are indicated below.

—Sinking shafts, driving levels, stoping, tramming, &c., 5000*l.*; dressing cost, timbering, banking, &c., 1600*l.*; enginemen, smiths, carpenters, sawyers, carters, &c., 1050*l.*; new plunger, jigger, shafting, launders, buddles, &c., 360*l.*; repairing workshops, cottages, stables, roads, &c., cleaning slime-pools, furniture, &c., 405*l.*; nine new horses (total number now 16), 455*l.*; new timber wagon, harness, &c., 66*l.*; iron tram wagons, new (now 103), 265*l.*; timber of all sorts, 400*l.*; iron and steel, 230*l.*; coals, 350*l.*; candles, 200*l.*; powder, 260*l.*; fuze, oil, grease and tallow, nails, bolts, ropes, shovels, 250*l.*; crusher rolls (new), 110*l.*; leather, belting, &c., 160*l.*; carriage of ore and materials, &c., 400*l.*: total, 11,561*l.* During the same period the ground opened, as will be seen by the agents' report, is as follows:—Sinking shafts, winzes, clearing, timbering, &c., 92 fms.; driving different levels, &c., 518 fms. 1ft.; stoping in different levels, 1107 fms. 4 ft.; making a total amount opened of 1717 fms.

In 1876 the venture reformed as the Great Dylife Lead Mining Company but with a depressed lead market there was little hope for a real recovery. In 1879 another new company took over and a considerable amount of work on all three lodes followed ; much optimism arose in connection with a 'new' vein called Alfred's lode (named after H. J. Alfred the managing director) which had been discovered in driving the great Dylife adit but had not been explored. A foot of solid galena was soon revealed and Alfred's shaft sunk from surface to 15 fathoms below adit. Unhappily the deposit proved of little substance and an attempt to sink the Llechwedd Ddu shaft below the 105 petered out due to lack of funds. The company collapsed in 1884.

*Dean remained manager until his death in 1881, aged 37. A white tombstone to his memory, and that of his young daughter Clara Minerva, may be seen in a corner of Machynlleth churchyard.

The account book of the period 1878-81 is preserved and provides a depressing narrative. Men employed gradually fell from 109 to 70 and average monthly earnings of the miners amounted to barely £3. But great mines take a long time to die, for however unfavourable the climate there are always investors gullible enough to risk capital in attempts to restore former glories.

From 1886 to 1891 Dylife restarted as Blaen Twymyn under the management of Evan Evans who resided in the nearby house of that name. Twenty or thirty men found employment and several hundred tons of lead and zinc ores were raised. Zinc blende abounded in the Esgairgaled lode but the hard nature of the veinstone and relatively low value of blende had hitherto restricted output. No copper ores were returned during this period, though previously the Llechwedd Ddu lode had yielded considerable tonnages of copper pyrites (often exhibiting the most beautiful iridescence).

About 1900 Evans laid a tramroad down the dingle from Esgairgaled engine shaft and across the road via a bridge to convey waste dumps to the dressing floors for treatment. This railway would have followed the same course as an earlier line to the floors.

The last activity took place in the late 1920's when Hirnant Minerals attempted to rework the enormous dumps below the road (since largely removed for hard core) which were optimistically estimated to contain 10,000 tons of lead. This failing, the company turned attention to re-opening the adjacent Dyfngwm mine and cleared out the Dylife adit with intentions of providing direct underground access. Subsequently, Alfred's shaft collapsed, completely blocking the adit so that nowadays the only entrance to the extensive workings under Dylife hill is by the process of laddering down the old shafts.

Whether mining will return to the area is doubtful, for in spite of all advances, the fact remains that lode deposits of this type are rarely so attractive today as in the past. Such were the achievements of the old men, of waterpower, horses and steam, and we are humbled by them.

Opposite. *An attempt to reopen Dylife in 1894. The prospectus, (which is full of inaccuracies), was promoted by the cousin of the Cardiganshire Matthew Francis. A family tree of this prolific family of mining agents is tabled in* The Old Metal Mines of Mid-Wales, Part 1.

DYLIFE MINE.

DEAR SIR,

This Mine is situated about 9 miles to the S. East of Machynlleth, in the County of Montgomery, and about 6 miles to the South of Llanbrynmair.

The formation is the Upper Silurian. The grant embraces over 600 acres, and it is a mile long on the course of the E. & West Lodes.

It is traversed by 5, E. & West Lodes, which naming them in the order of occurrence from North to South are as follows :—

1. Esgirgaled.
2. Llechwedd-du.
3. Intermediate Lode, dipping North (New).
4.   Do.   do.   do.   South (New).
5. Dylife, or Dyfngwm Lode.

Only 3 Lodes have hitherto been worked, viz., 1, 2 and 5, and each proved remarkably rich the profits in the aggregate, having been more than a quarter of a million pounds sterling. In one year the balance to the credit of the company which formerly worked it, with which the late Messrs Cobden & Bright, were closely associated exceeded £20000. The monthly returns of Lead Ore averaging 300 tons. In the Llechwedd-du Lode, No. 2, there are nice runs of Ore going down below the bottom level, 300 yards deep, and the deeper development of this, as well as of the Dylife and Esgirgaled Lodes, offers a fair chance of success. But the greatest importance is attached to the results that are expected to attend the development of the Virgin Lodes, 3 and 4, situated between the Llechwedd-du and Dylife Lodes, which can be undertaken and carried out at a comparative small cost, by driving a crosscut at a depth of 70 fms. exploring them therefrom. These Lodes 3 and 4, are well defined on surface and each shows a little Ore, looking quite as well as the other Lodes which proved so rich, and in my opinion there can scarcely be a doubt that if developed in depth they will prove very rich. A corresponding trial at the " Van Mine," situated in the same Geological formation, about 6 miles to the S. East, has resulted in a great success. A crosscut has been driven South, and a New Lode has been discovered which is very rich.

The Mine is thoroughly equipped with Pumping, Winding, and Dressing Machinery of a Modern description, all worked by Waterpower, the cost of which would be several thousands of pounds. It has capacious Workshops supplied with good tools, and there are plenty of Rails and Materials which will suffice for all requirements for a considerable period. The Annual Rent for the Waterpower is only £6, and the Royalty but $\frac{1}{25}$ th. with a redeemable dead rent of only £10 a year.

The standing charges for keeping the Mine drained and in working order are very small, and inasmuch as the length of the crosscut would not exceed 200 yards, to intersect the two Lodes which can be driven at £3 a yard. The cost of the important trial indicated would not exceed £1400, after allowing a fair sum for the exploration of the Lodes after they have been intersected.

I propose the formation of a syndicate holding shares in multiples of £100 each, to raise £2000, and to show my confidence in the property, I am prepared to take one share.

I have great confidence that before twelve months elapse, a substantial discovery would be made, and it is probable that ere then there will be improvement in the price of Lead, either of which happening would admit of a formation of " Limited Liability Co.," to take the property over at a substantial profit to the syndicate.

I have visited the property on three occasions and have incurred considerable expense in making myself acquainted with the facts of the case, but I do not ask to be recouped, and I shall be quite content to join the syndicate, on the same terms as are conceded to others.

I am, Dear Sir,

Yours truly,

MATTHEW FRANCIS.

(Signed)

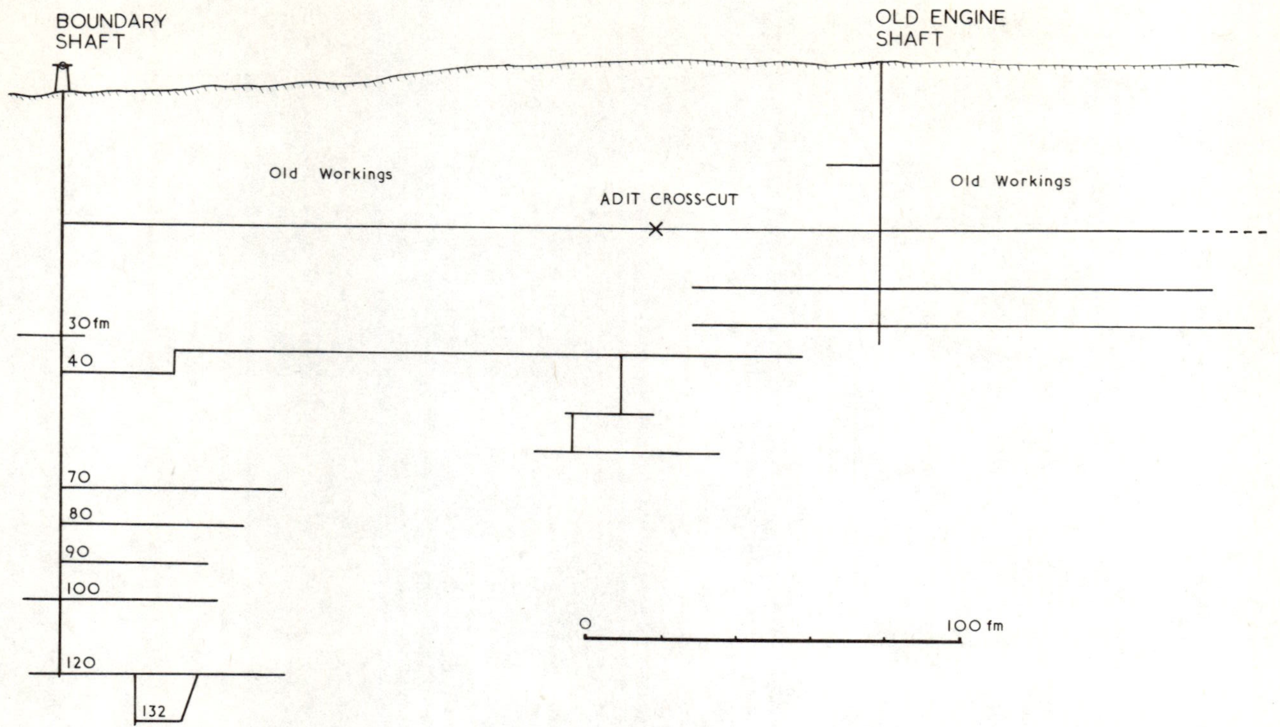

Dylife lode. *A section of the workings from an original plan dated 1877, deposited in the Mining Record Office, London. There are extensive stopes above adit level.*

MACHINERY AND INDUSTRIAL ARCHAEOLOGY

This chapter will cover the industrial archaeology of the mine and will also include notes on the machinery employed, the two being extensively interlinked.

Perhaps the oldest surviving plan is dated 1774 'An account of the Lead Mine in Delivie Mountain,' and is of particular interest in showing a waterwheel for driving pumping machinery. As to the location, the workings are described as 'on a Hill called Esgir Galed', and no doubt on the Esgairgaled lode, but whether the stream referred to is Afon Twymyn or Engine Dingle, is hardly possible to decide. The latter, however, seems the more probable. The plan is reproduced below.

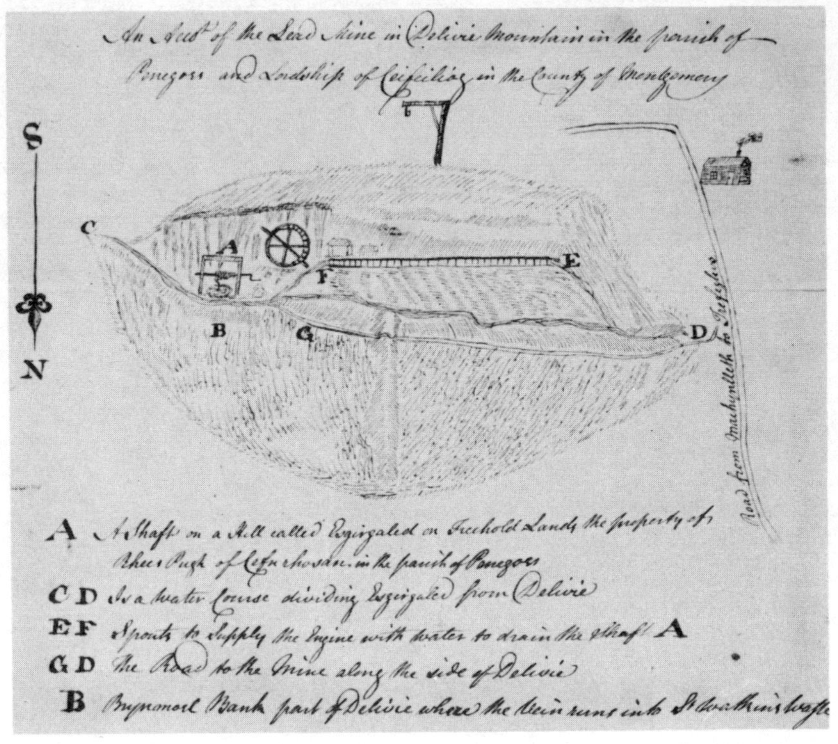

N.L.W.

A waterwheel for pumping was certainly operating in Engine Dingle during the Williams and Pughe period, on the north side of the stream and close to Esgairgaled engine shaft. It also pumped Llechwedd Ddu shaft by means of a long line of flat rods passing in front of Esgairgaled cottage.*This 'beautiful and costly' machinery commenced in motion on 7th June 1839 and Welsh verse was 'prepared for the occasion supposed to be spoken alternately by the Engine and the daughter of Mr. Pughe'. Nowadays of course, the idea of any form of conversation with pumping equipment is too ludicrous to contemplate, but its reciprocating ancestor making only three or four strokes every minute, presented the opportunity for an impressive dialogue matched to the scale and dignity of the machinery itself.

At the time of Francis's report (1849) the machinery consisted of a 37ft wheel for pumping—probably the one just mentioned—and an 18ft wheel for the crushing mill at dressing floors near the confluence of the two streams. Drawing was accomplished by horse whims. Francis advised much larger and more powerful waterwheels, the construction of several reservoirs, and tramroads to convey ore to the mill.

Two years later Dylife was 'stocking with machinery of a gigantic character' and waterwheels of 40ft and 63ft diameter were being erected. A traveller in 1853 was 'struck by a large and long embankment empounding a most beautiful sheet of water adjoining the main road' (Rhyd Y Porthmyn Pool). A reservoir 54ft high and several hundred feet long was under construction and also a row of 20 cottages, known as Rhanc Y Mynydd. The reservoir (Llyn Nant-ddeiliog) in a remote valley to the north, necessitated a long leat and its forgotten waters are the worthy object of a walk over the hills. The two ponds in the headwaters of Engine Dingle had probably been built by this time ; all these reservoirs are clearly shown on Ordnance maps.

The development and extension of waterpower brought considerable changes. The line of flat rods from the Esgairgaled wheel gave way to the new 63ft×3½ft wheel, called the Red Wheel, providing power for drawing and pumping the Llechwedd Ddu shaft. This wheel was, I believe, the largest ever erected in Wales and must have presented a magnificent spectacle under full load. Near the side of the road a wheel known as the Black Wheel was installed for pumping and drawing at Esgairgaled shaft and according to Will Richards was the first on the mine—no doubt none

*Waterwheels driving reciprocating pumps required delicate adjustment, especially when in conjunction with long runs of flat rods. Balance boxes were usually necessary at both the shaft (to counteract the weight of pump rods) and at the wheel itself. Overall efficiency was about 50% leading to a simple and useful formula thus 2×height of pumping×outlet flow= wheel diameter×flow over wheel. The application of waterpower to mining purposes is described in some detail in *The Old Metal Mines of Mid-Wales Part 1.*

Will Richards

This old postcard dates from about 1908. On the left is John Lloyd, for many years blacksmith at Dylife. In spite of its appearance the building could not have been more than 60 or 70 years old.

21

Above. *The massive iron gudgeon that supported the balance bob at Boundary shaft.*
Opposite. *The erstwhile majesty of the Red Wheel is indicated by its pit, now sadly employed as a rubbish tip. The pump rod at Llechwedd Ddu shaft can be seen in the distance.*

other than the old 1839 wheel re-sited to take advantage of the new leat system. It should be mentioned that all these modifications were carried out under the direction of Captain Edward Williams who was one of the most important and respected personalities of the Williams and Pughe era. He managed the mine for 26 years, before leaving in the 1860's to become agent to several other mines in the vicinity.

Dylife was thus well equipped by the standards of the day when taken over by Bright & Co. in 1858. Nevertheless as we have seen, improvements immediately commenced, especially concerning the drawing machinery. A double-decked cage with Aytoun's patent safety catches were fitted to Llechwedd Ddu engine shaft and operated by an inch diameter steel wire with a breaking strain of 27 tons.

As the mine developed, cages were also fitted to Bradford's shaft which eventually reached a depth of 130 fathoms, or 25 fathoms deeper than Llechwedd Ddu shaft. To ease the load on the waterwheel, the drawing

machine comprised one drum for each shaft so arranged that one cage descended as the other rose. The two drums had different diameters to allow for the respective depths.

Power was obtained by applying water to the wheel via a cable operated from the drawing machine house which closed a sluice in the launder. Controlling such massive machinery, which always ran unevenly due to the cyclical pumping load, must have called for great experience and dexterity. In the 1870's this task was carried out by John Hughes who also operated the Boundary shaft drawing machine as described below. He seems to have been the engineer to the mine and lived at Esgairgaled, now an inconspicuous farm building, long since abandoned for domestic purposes.

The Star Inn at the time of the author's visit in 1950. It is now modernized, with items of mining interest displayed including a fine portrait on slate of Captain Edward Williams (see opposite).

The mine trams were 3ft6in×2ft6in×20in deep and could also be drawn to surface at Esgairgaled shaft, though this utilized only a single decker cage. The latter survived *in situ* until very recently and one or two pieces of iron may yet be seen at the shaft collar.

At Boundary or Waller's shaft on the Dylife lode, developments of a different nature took place. On this windswept plateau a 60in Cornish pumping engine of 9ft equal beam was installed together with two 10 ton boilers and held an unenviable record for the most elevated (1450ft O.D.) engine of its size in Wales. Hauling the heavy components over the severe grades from Llanidloes is said to have demanded the efforts of 50 horses.

The cost and transport of coal proved to be a crippling burden and it was later claimed that waterpower could have accomplished the task—if less reliably. Presumably such means of pumping had previously served Old Engine shaft, though we are unfortunately at a loss for any details and even the location of the wheel is obscure.

This understandable wariness of steam led in 1861 to the expedient of erecting a 50ft×6ft waterwheel no less than a mile away for drawing at both Old Engine and Boundary shafts. Such a feat was made possible by the introduction of wire rope, which was much stronger for a given weight than hemp. The site was just below the road about 500 yards SSE of the Star Inn and the massive wheelpit is well worth examination. It is carved out of solid rock and the location of the drawing machine can be seen alongside.

24

From this machine a cable ran right over Dylife mountain (see pg. 4) and its route is marked by several old posts that no doubt held the rollers. The cable passed beneath the old coach road in a gully and culvert, still evident, about 200 yards east of Penycrogbren* ; the whole must have formed the longest and most powerful water operated drawing equipment in Britain. Frost and thermal expansion would have presented problems, and with only a wire rope and knocker the difficulty of signalling in all weathers across a mile of moorland beggars the imagination. Nevertheless Mr. Richards assures me this was the method, and adds that the local postman earned a shilling or two by oiling the rollers. For drawing from Boundary shaft an extra length of rope was added on to the main cable at Old Engine shaft.†

*Penycrogben (Gallows hill) derives its name from gallows that stood here, as depicted on page 19. About 1700 a blacksmith at the mines murdered his wife and disposed of the body down a nearby shaft. Some years ago, Will Richards unearthed at the gallows site an iron headpiece holding the skull of the murderer—this awesome relic is now in the National Museum of Wales.

†In the mid-nineteenth century, distances between shafts and drawing machines frequently exceeded 1000 yards at the lead mines of Grassington in Yorkshire ; however 1300 yards seems to have been about the limit in that district.

Steam also found employment for standby purposes at the Red Wheel and on the dressing floors, where two high pressure 'puffers' were working in 1865. By 1874, a 20in engine with two 6 ton boilers aided the Red Wheel and there was a 12in engine with 5 ton boiler for dressing. On the floors were two large waterwheels, one of which drove two crushers, a stone-breaker and jiggers. The other operated six round buddles, a Zennor buddle and other machinery.

As already mentioned, considerable work afterwards followed on the Dylife lode, and a balance bob was fitted at the 120 fathom level to ease the load on the pitwork. Underground bobs were a rarity in Wales and in this case hardly justified, as sinking only proceeded a further 12 fathoms. This however made a total depth of about 167 fathoms—the deepest shaft in Mid-Wales.

With regard to the Cornish engine, a few details survive concerning day to day running ; for instance in the four weeks ending 7th September 1878 the engine house was cleaned and whitewashed and $4\frac{1}{2}$ days were spent packing the piston. Work of this nature was usually done by two brothers William and Henry Goldsworthy. The engine ceased work in January 1880 and appears to have lain many years derelict before being scrapped. Unfortunately, as is so often the case in respect of Welsh mines, no contemporary illustrations are known, and little except the foundations remain of the building. There is however, a large grass-grown cinder tip which testifies to the amount of work done.

Towards the end of the nineteenth century, the outer portions of the wooden spokes of the Red Wheel were becoming decayed and to avoid the costs of renewal, the expedient was adopted of reducing the diameter to 60ft. This modification presumably explains the 'step' in the launder, as seen on the front cover.

Most of the plant and equipment was finally sold about 1912 but in so remote a locality many items would not bear the cost of removal, amongst these being the Esgairgaled and Llechwedd Ddu shaft cages. By some means the latter became removed some distance downstream, where it lay buried until 1970 when I stumbled across a portion of its ironwork protruding from the bottom of a tip. This cage, which may be said to epitomise Dylife in its greatest days, is now preserved at the Mid-Wales Mining Museum, Ponterwyd. Where restoration on site is not practical, museums and other places of display accessible to the public are preferable to the oblivion of private collections for relics such as this, and it is pleasing to record that various items of local mining interest are on view at the Star Inn.

Cast iron railway chairs for 'bar' rails (an early form of permanent way) have been found in the Llechwedd Ddu area, and other discoveries are sure to await the industrial archaeologist. He will find nowhere better to probe and ponder than amongst these glorious hills.

Top. *Footway shaft, giving access to the Llechwedd Ddu lode. This probably dates from the 1820s. (see pg. 10, item 7).*
Bottom. *Industrial Archaeology in practice—revealing the Llechwedd Ddu cage.*

John Bright, M.P. 1811-1889

Opposite. An approximate survey of adit workings, made in 1970. The bold lines indicate drivages on lodes.
E=Esgairgaled Lode   L.D.=Llechwedd Ddu lode.

1   *This level emerges close against the Afon Twymyn east of the 50ft waterwheel pit, and was driven by a Captain Reynolds in an effort to locate the Dylife lode.*

2   *Pencerig deep adit.*

3   *Pencerig shallow adit which is connected to the deep adit by stopes. It is also stoped out to surface—the floor of opencast workings visible above rests on timbers which will ultimately collapse.*

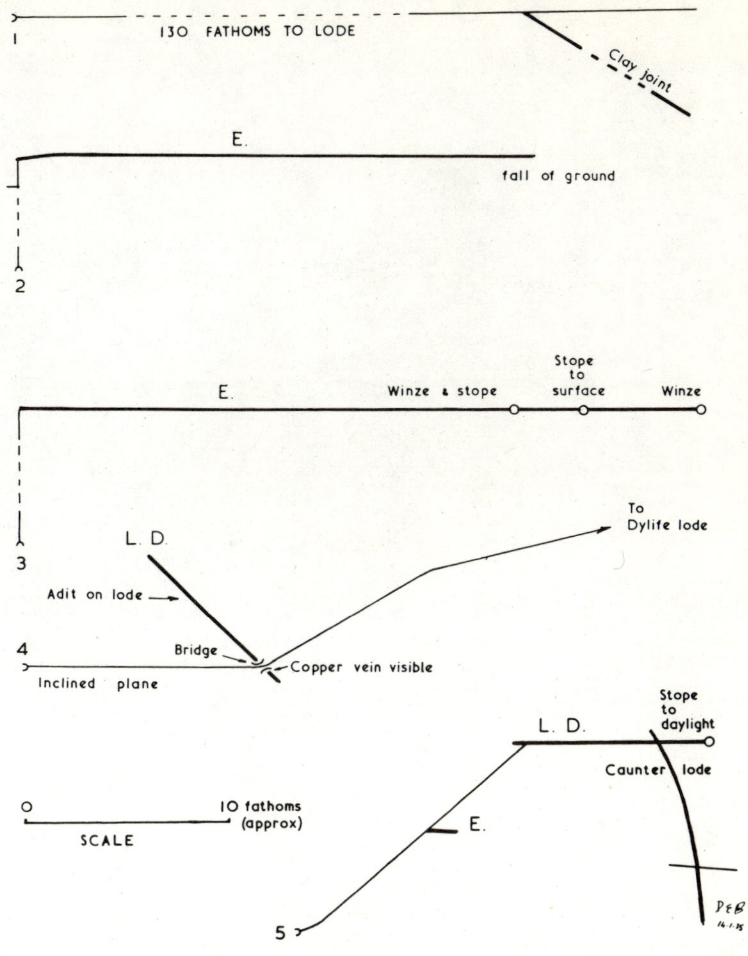

4   *Inclined plane and cross-cut adit to Dylife lode. Opposite this level the Gwaith Gwyn adit is no longer accessible, having been silted up by debris from the stream.*

5   *Level Goch. The mouth is now almost obscured. This area is of interest, being the intersection of the two lodes, where workings extend up to daylight and very probably to a considerable depth. The old miners believed that the lodes did not unite, but crossed, and extensive surface trials were made in consequence.*

OUTPUT STATISTICS
From commencement of official returns.

Company	Year	Lead Ore (tons)	Copper Ore (tons)	Company	Year	Lead Ore (tons)	Copper Ore (tons)	Zinc Ore (tons)
Williams & Pughe	1845	586	?	Cobden & Bright	1872	932	126	30
,,	1846	557		Dylife Lead Mining Co.	1873	782		
,,	1847	524		,,	1874	820	31	30
,,	1848	545		,,	1875	1196		
,,	1849	588		Great Dylife Companies	1876	1002	36	
,,	1850	653		,,	1877	860	26	
,,	1851	1003		,,	1878	636	21	
,,	1852	750	72	,,	1879	527		
,,	1853	705		,,	1880	315	53	
,,	1854	637	22	,,	1881	325		
,,	1855	192	28	,,	1882	430		
,,	1856	489	43	,,	1883	334		
,,	1857	828	83	,,	1884	90		
Cobden & Bright	1858	318	110	—	1885	—		
,,	1859	1200		Blaen Twymyn	1886	—		
,,	1860	1267		,,	1887	60		60
,,	1861	1198	115	,,	1888	18		130
,,	1862	2571		,,	1889	145		29
,,	1863	1714		,,	1890	167		41
,,	1864	1689		,,	1891	67		11
,,	1865	1537		Dylife Mining Co.	1892	61		20
,,	1866	1304		Sir W. W. Wynn	1893	6		
,,	1867	2161						
,,	1868	1419	432	Evan Evans	1899	16		24
,,	1869	1448	155	,,	1900	8		14
,,	1870	1028	120	,,	1901	1		2
,,	1871	973	67					

Total ore tonnages since 1845: *Lead,* 36,684; *Copper,* 1,540; *Zinc,* 391.
(*Silver,* about 160,000 oz.)

SOURCES AND ACKNOWLEDGEMENTS

Information relating to mines such as Dylife which have worked on and off for centuries is bound to be widely scattered, so that to compile a complete list of source material is virtually impossible. The following references have provided the main sources for this book.

The chief repository of primary material relating to Dylife (and most other metal mines in the principality) is the National Library of Wales. Here are the Wynnstay Estate documents and various estate maps together with the Druid Inn collection. The latter comprises personal papers of the late Matthew Francis and is of the utmost value relative to the 19th century mining scene and its motivation.

At the British Library, the Cobden and Bright papers have been consulted. The task is laborious, but no doubt a thorough search would throw further new light on the period.

Of published sources the *Mining Journal* is quite the most important, though unfortunately its columns are reticent about the private company periods prior to 1873. Though Richard Cobden and John Bright have formed the subject of numerous biographies, I have been unable to find the least reference to Dylife—an unexpected omission. The mineralogist will find early notes on the minerals of Dylife in J. Woodward, *A Natural History of the Fossils of England,* 1728, ii. 27-30.

For the economic geology of the area O. T. Jones *The Mining District of North Cardiganshire & West Montgomeryshire,* 1922, is to be consulted, and the standard introduction to the industry's history is W. J. Lewis *Lead Mining in Wales,* 1967.

This volume has been largely inspired by the enthusiasm and reminiscences of Will Richards, a lead miner in his younger days, whose lifetime's love and knowledge of Dylife have contributed much to its pages. To him my best thanks are due. I am also indebted to George W. Hall for many extracts from the *Mining Journal*, to Trevor Morris, and to the staff of the National Library of Wales and the British Museum for help on various occasions.

Mr. Leslie W. Anthony of the Star Inn and Mr. D. Wilson, Blaen Twymyn, have kindly permitted reproduction of the illustrations on pages 25 and 17 respectively; not least I have to thank my wife for typing the manuscript and assisting with surveys in the field.

GLOSSARY

ADIT	A tunnel driven into a hillside for access and drainage.
BLENDE	The common ore of zinc, usually with a semi-metallic blue/brown lustre. Contains up to 67% zinc, 33% sulphur, plus iron as an impurity.
BUDDLE	A machine for separating finely crushed ore from gangue.
CROSS-CUT	A level or adit driven in order to cut a lode.
DRAWING MACHINE	The old term for a winding engine.
DRESSING	The process of removing waste material and gangue from the crude or ' undressed ' ore.
ENGINE SHAFT	A shaft fitted with pumping equipment.
FLAT RODS	Lengths of iron bar about 20 ft. long, pinned together and supported on rollers or jockey wheels, for the purpose of transmitting power.
GALENA	The common ore of lead, distinguished by its great density and silvery metallic lustre when freshly broken. The external colour is light grey after long exposure. 86% lead, 14% sulphur.
GANGUE	The ore matrix, often comprising fragments of rock cemented with quartz, calcite, etc.
GOSSAN	The weathered upper part of a lode.
KIBBLE	A barrel-shaped vessel of iron or wood in which ore was raised.
LAUNDER	A channel for conveying water, usually made of wood.
LEVEL	A tunnel or gallery, usually on the course of a lode.
LODE or VEIN	Mineralized ground containing ore usually in company with valueless material, e.g. rock, quartz, etc. (see Gangue). Lodes are often nearly vertical, 20 or 30 feet wide, and can extend for miles.
OPENCUTS	Quarry-like excavations on the lode, often long, narrow and deep, and dating from ancient times.
SKIP	A rectangular iron vessel with four wheels guided by a wooden track fitted into the shaft. This device eventually displaced the kibble.
STOPE	The working chambers or cavities in a mine created by removal of ore.
WINZE	An underground shaft excavated downwards.